West Asia Tours of Duty
Combat Tested

Vickey Joins the Military Series

Book 7

West Asia Tours of Duty
Combat Tested

Book Seven

Table of Contents

Road to West Asia, the Combat Tours 4

First West Asia Tour.. 14

Second West Asia Tour 28

Third West Asia Tour 33

Road to West Asia, the Combat Tours

First, becoming a Warrant Officer.

Reflecting on the past, it was clear that my journey to combat began long ago, as I geared up for a life dedicated to service. As a graduate of the Barbizon School of Modeling and a recipient of the Ms. Congeniality award—well before the film of the same name was released—I initially trained for service in the news industry. Everything leading me down that path was unfolding as planned until a different call to service captured my attention and it was that call that led me down the path to becoming a combat tested Warrant Officer.

During my tenure as a Warrant Officer on multiple tours in West Asia, it became apparent that the tale of that journey, becoming and being a Warrant Officer, became significant and needed to be highlighted. The path to becoming an air defense warrant officer involved traversing various roads, all of which led to combat tours.

Joining the Army later in life, I endured boot camp, advanced individual training, and various duty stations and roles before finally reaching the air defense technical warrant officer school. I soon realized that graduating from the Warrant Officer Candidate School was merely the beginning of my transformation from soldier to Warrant Officer.

Upon completion of the candidate school, every Warrant Officer, irrespective of their specialty, was bestowed with the warrant officer eagle. The emblem, representing America, the American soldier, and exceptional achievement, instantly evoked images of professionalism, reliability, dedication, competence, and technical expertise.

Soldiers often held the belief that warrant officers knew everything. They believed that warrant officers were gifted with a sixth sense that enabled them to know instantly what was wrong with the equipment. So, whenever something malfunctioned, the immediate response was clear: consult "Chief," the officer wearing the eagle.

The Warrant Officer insignia!

Many were under the impression that warrant officers could rectify equipment issues merely by their presence or by simply touching the machinery. While the notion appeared absurd, surprisingly, it often seemed to unfold just so.

However, before I could attain that revered status, I needed to complete the Warrant Officer Basic Course. The program was designed to certify my technical and tactical competence as an air defense technical expert. Throughout the course, we learned the fundamental principles of becoming an effective officer, which included decision-making lessons, briefing techniques, and, crucially, the technical specifics of our roles.

As graduation from the Warrant Officer Basic School neared after nine months of rigorous training, a wave of panic set in. Much like at the end of high school, I was on the verge of stepping out from the protective environment of an educational institution into the high-stakes world of air defense. No longer would I have instructors guiding me; I was about to become “the expert” that others depended on to resolve issues.

Once I graduated, I was officially certified for my first duty assignment as a newly appointed Warrant Officer. The weight of responsibility as a unit warrant officer began to press down on me. Unlike more seasoned warrant officers, I had not yet accrued extensive field experience or spent long hours ensuring a unit was mission-ready. No soldiers had called me in the middle of the night with equipment failures, nor had a commander relied on me for the success of a unit operation. The mounting stress made me question whether I was truly ready for the challenges ahead.

As I arrived at my first duty station as a Warrant Officer, my fears of failure intensified. Up to that point in my career, I had successfully navigated through each stage of my career advancement and job responsibilities. However, fully preparing for the role of a technically and tactically proficient warrant officer was a different challenge. In the maintenance and operations domain of air defense, no two problems were identical, which meant there was always a learning curve. As a Warrant Officer, expectations were high; I was to be the best of the best, the technical expert, "Chief." The pressure escalated daily as I strived to master my new role.

The responsibilities of being a Warrant Officer turned out to be far more extensive than I had anticipated or been briefed on. The role extended beyond merely operating or maintaining equipment. I was tasked with maintaining reports, navigating supply channels, and briefing commanders on the unit's capabilities and readiness status, among other duties. Those were just a few of the additional responsibilities that came with being a unit warrant officer.

I was initially unaware of all the requirements, details, and expectations associated with the position of a Warrant Officer. I hadn't witnessed all the behind-the-scenes efforts that my "Chief" in Korea undertook. I was quickly learning why everyone believed "Chief" had all the answers: it was because we were expected to have them.

So, the panic was always there waiting to poke its head out if I could not arrive at the solution to a technical problem, or my reports were erroneous, or I failed to have the answers during a briefing. That sense of panic intensified during high-pressure situations, which were frequent. Until I could demonstrate my ability to resolve complex issues under such pressure, I felt like merely a bearer of the warrant officer insignia and not a true warrant officer. Failing in my duties would not only be a personal defeat but could also tarnish the reputation of the Warrant Officer Corps. That was a heavy load to carry, particularly for someone straight out of training.

Upon my arrival at the new unit, it was standard procedure for a meeting to be arranged with the Commander. During my meeting, the Commander outlined the operational status of the unit and emphasized my crucial role in maintaining our readiness. However, I immediately felt that the Commander harbored doubts about my capabilities. There were no words of encouragement; instead, his demeanor suggested that he had written me off as contributing little to his command.

I had encountered challenges before, but that was an unprecedented level of overt skepticism towards my abilities. It was at that point that I resolved to prove the Commander wrong. His dismissive attitude not only failed to deter me but actually ignited a stronger resolve within me. Adopting the mantra “failure is not an option,” I was fueled by a determination to not only meet but exceed expectations, turning skepticism into respect.

Determined to excel, I immersed myself in learning everything about my new role and absorbed as much knowledge as possible. I knew that eventually, I would have the chance to demonstrate my capabilities to myself, my soldiers, and my commander. With time on the job and guidance from senior Warrant Officers, I started to devise innovative ways to troubleshoot and repair downed systems, even developing techniques that were not outlined in the technical manuals. Though I was gaining proficiency, I had yet to fully earn the trust and confidence of my unit.

I continued to collaborate closely with my crew, learning from each other and collectively finding solutions to various challenges. As my skills and confidence grew, so did my ability to lead effectively. It was during my unit’s deployment to Saudi Arabia that I truly proved my mettle and earned my eagle wings, demonstrating that I could rise to the occasion and lead under challenging circumstances.

One evening, as I arrived for my shift, I was greeted by a system that had been non-operational for an entire day. Left on site to tackle an issue that more seasoned experts had failed to resolve, I initially felt quite intimidated.

Once the prior shift personnel had cleared out, I took the opportunity to thoroughly review the manuals and reapproach the problem from scratch. I considered similar past incidents and then deliberately stepped back to reflect. It was during that reflection time that I arrived at the possible solution. Confident in my analysis, I coordinated with my team to implement the fix. Together, we successfully brought the weapon system back to full operational status, a moment that underscored my growing competence and leadership in the field.

That successful repair marked a turning point, elevating my status to that of a certified warrant officer within the Warrant Officer Corps. News of the fix quickly spread among other crews, units, and importantly, reached the Commander. Seasoned system maintainers, previously baffled by the problem, sought me out, eager to learn how I had managed the repair. I was pleased to share my knowledge and contribute to our collective expertise.

The Commander acknowledged my achievement with a formal commendation, reflecting a significant shift in his perception of my capabilities. That solution not only bolstered my confidence but also earned me the respect of my peers and superiors. The respect and recognition culminated in the honor of being tasked with reenlisting members of my unit, affirming my leadership and valued role within the Army. The series of events firmly established my position and future within the military, setting the stage for continued success and influence.

What an honor and privilege to reenlist a soldier!

My journey to becoming a respected Warrant Officer was anything but smooth. It required a cycle of listening, learning, performing, and facing failures before I truly earned the respect that accompanied the warrant officer's insignia. My initial assignment as a Warrant Officer laid the critical foundation necessary for mastering my responsibilities.

That duty assignment not only took me to Saudi Arabia for my first combat tour but also coincided with a period of profound national tragedy. It was during that rotation that the events of September 11, 2001, unfolded, forever altering the landscape of our nation and deeply affecting its people.

First West Asia Tour

Saudi Arabia (2001)

My assignment in Saudi Arabia started in June of 2001 at Riyadh Air Base, where I was stationed at Eskan Village. Just a day after my arrival, the military escalated to a critical threat condition posture due to heightened regional tensions. That required all personnel, including myself, to wear body armor and have our biological gear readily accessible at all times. That immediate shift to a high-alert status underscored the volatile environment in which we were operating, setting the tone for a demanding and cautious deployment.

Saudi gear, body armor and abaya!

Within just twenty days of arriving in Saudi Arabia, we experienced a missile launch alert—a stark initiation into the realities of a combat tour. The event was a vivid reminder of the constant dangers present in the region.

A month into my combat responsibilities, another significant incident occurred. While en route to the weapon system site, my driver and I were stopped and detained at gunpoint by Saudi security guards. The encounter was alarming, and it hinted at the complexities and challenges that lay ahead. Even before the detention, there had been several indications that my tour of duty in Saudi would be marked by unusual and potentially perilous events, setting a tone of unpredictability and vigilance for the duration of my stay.

Upon my arrival at the base, the initial routines of gearing up unfolded as expected. Receiving biological gear and body armor was anticipated, given the combat zone setting. However, the issuance of the abaya, a robe-like dress worn by women in the region, was an unexpected directive, especially since we were US service members.

The requirement to wear the abaya signified a deeper level of cultural integration than many of us anticipated. Despite being American, we were expected to respect and conform to the local customs. The requirement was a clear indication of the complexities involved in operating within a foreign and culturally distinct environment, highlighting the delicate balance between military operations and cultural sensitivity.

American military service women!

Navigating the regional customs added significant layers of complexity to our military operations. One such custom dictated that women, regardless of their rank or role, must ride in the back seat of vehicles. As a senior ranking officer, that was particularly challenging for me because it contradicted the military norm where I would typically sit in the front as the commanding officer.

Moreover, interactions with Saudi males presented further difficulties. They did not acknowledge American women, regardless of their rank, which meant they refused to take instructions from me or even engage in basic communication. That disregard was deeply problematic, as was starkly illustrated during the gunpoint incident.

During that tense situation, my ability to command or protect my soldier was severely undermined. My driver, mistaken for a foreign national, faced aggressive and frightening treatment. He was forced out of the vehicle, made to lie on the ground, and was jabbed in the back with rifles while being threatened with shooting. Witnessing him in such distress, with tears in his eyes, was profoundly distressing. It was, without doubt, the most tense and horrifying experience I had encountered in my military career up to that point. The combination of being powerless to assist and the harsh enforcement of local customs made it an extraordinarily challenging ordeal.

Amid the confrontation, I frantically tried to communicate our identities and purpose, shouting over the din to protect my soldier, but my attempts were futile. To the Saudi security guards, I was virtually invisible, and my explanations and pleas were blatantly ignored. In desperation, I reached out to my unit for assistance, but help was slow to arrive, leaving us in a precarious situation for what felt like an eternity.

The ordeal dragged on for hours under the bleak desert sky. Eventually, after exhaustive efforts to confirm our identities and validate our authorization to be in the area, we were finally released to continue to our destination. The relief of being let go did little to mitigate the deep fear and lasting trauma inflicted by such a harrowing experience.

As our tour of duty progressed, challenges continued unabated. The very next month, we were again placed under a critical threat condition due to renewed missile threats. That time, however, we were not merely passive recipients of alerts; we were actively engaged in countering the threats. The month culminated with the US military maintaining a heightened state of alert, perpetuating the constant stress and vigilance that defined our deployment.

As the tour progressed, each duty day started in a routine fashion: an early rise followed by preparation for the day's tasks. My crew and I would load up the van and prepare for travel, and I would don my travel attire—armed with a loaded weapon, wearing body armor, and clad in the abaya, in accordance with the customs of Saudi Arabia. That had become our standard procedure during the last four months of our deployment in the Kingdom.

Despite my rank and role as the leader, local customs dictated that I took my seat in the rear of the vehicle, as was required for all women in the country, regardless of their status or profession. Once settled, the team and I would engage in our usual mix of light-hearted jokes and storytelling. That camaraderie helped to lighten the atmosphere and maintain morale amid the strict protocols and the tense security environment we navigated daily.

But on one day, the day would end with an unusual turn of events. But the start of the day began as most other days. Upon reaching the work site, the day unfolded smoothly. The missile system was fully operational, poised for action at a moment's notice, and all logistical operations were executed flawlessly. With our primary tasks handled efficiently, we found time to enjoy a few rounds of cards, which served as a welcomed diversion and helped us unwind.

As the duty day wrapped up, we packed the van and prepared for the hour-long journey back to our living quarters. The ride back was filled with the usual exchange of jokes and tales, a light-hearted end to an uneventful day. However, unbeknownst to us, we were returning to a situation far from ordinary.

As we approached the entrance of our base, we noticed the guards were operating under a heightened threat condition—a sight familiar yet infrequent enough to cause concern.

"Oh man, what has some knucklehead done now," quipped Specialist Garcia, capturing the mix of annoyance and tension that came with such disruptions. The atmosphere shifted from the casual banter of our journey as we braced for whatever awaited us at the base.

Though not a member of my system maintenance team, Garcia served as the parts clerk within our shift's crew, also doubling as my driver whenever travel was required.

"Man, it's just another pointless exercise," remarked Staff Sergeant Gaines, our system maintenance noncommissioned officer. "These threat condition drills are becoming absurd."

"Don't fret, gentlemen, there's still ample time to make it to the pub for drinks," I half-joked, half-worriedly. "Let's have those IDs ready to expedite the process." As we approached the entrance, my concern grew slightly, sensing that the event might deviate from the norm.

Approaching the guards, Staff Sergeant Gaines inquired, "What's happening, buddy?" as he presented his ID, recognizing the guard.

"Man, the World Trade Center's been bombed," the guard replied, visibly shaken and speaking with urgency.

"Man, you shittin' me?" Gaines responding with shock.

"Nah dude, it is all over the monitors in the orderly room," the guard stated with utter shock in his voice. "Man, this shit is serious."

After successfully navigating the checks, Garcia promptly parked the vehicle, and we dashed to the orderly room, eager to witness and hear the news firsthand. And indeed, it was all true. We stood there, stunned and astonished, as we watched the unfolding events on the monitor. While we were stationed on foreign soil, striving to maintain security and act as a deterrent to escalating tensions, an attack had transpired on American soil.

Due to the time disparity between the United States and Saudi Arabia, we were witnessing the events well after their occurrence. Yet, as my team and I observed the events replaying on the monitors, it felt as though they had just taken place.

Suddenly, approximately 15 minutes after our return to base and absorbing the news of the World Trade Center attack, the blare of the base alert siren pierced the air, prompting all base personnel to scramble into formation. That was a routine protocol for alerts - formations were organized swiftly to ensure accountability of all personnel.

Still donned in our gear, my crew and I hurried to our designated formation area without delay.

"Did you guys see what happened?" Chief Warrant Officer Two Joseph asked me.

"Yes," I stated.

The commander approached the formation to deliver the news that the base had been elevated to a heightened DEFCON or defense condition. We were now at DEFCON 2, a significant escalation from DEFCON 5, signifying an increase in force readiness, just shy of the maximum readiness level of DEFCON 1.

Thus, on September 11th, we found ourselves once more under heightened threat conditions. At that moment, we were directed to keep our biological gear within reach, verify the integrity of our masks, and ensure our canteens were filled. That precautionary measure was in response to the attacks on the World Trade Center and the Pentagon in America.

As the United States commenced its response to the terror attack, the conflict was initially dubbed "Operation Infinite Justice," later renamed "Operation Enduring Freedom."

For myself and my unit, the onset of war, known as Operation Iraqi Freedom, was marked by the events of September 11, 2001, when terrorism struck the World Trade Center in New York City. Consequently, my tour of duty in Riyadh, Saudi Arabia, took an unforeseen turn.

Our task force, originally designated Task Force Southern Watch, underwent a transformation into Operation Enduring Freedom. What was once a routine mission had become a real-world event, stirring high emotions among those of us stationed in Saudi Arabia. There was a palpable desire for immediate action, but fate had other plans for us.

Following the commander's announcement, I convened our entire crew to outline our new operational protocols. We would persist with our round-the-clock operations and maintain our shifts, but with a significant alteration – we would all remain on-site, refraining from returning to base until the threat level subsided.

"Get your ready bags and anything else you think you might need for a week-long stay on site," I instructed the crew.

"Chief, do you think our tour will be extended here in Saudi?" Garcia asked.

"I don't know Garcia, but it is definitely a possibility," I briefed him.

As the crew dispersed to collect their gear, I made my way to speak with the Commander.

"Sir, my team and I will be heading back to site and we'll stay put until further notice. Do you believe our tour will be extended?" I asked.

"Well Chief, it is a strong possibility. I expect a brief from the Colonel shortly and I'll let you and your guys know what's happening," Captain Brown stated.

"Okay sir, we're headed to site."

Time elapsed, and the threat level gradually decreased, with no immediate response to the attack. However, on October 6th, we found ourselves once again on high alert. We received word that the

US was preparing to strike the enemy within 48 hours. Initially, we were alerted through the television, prompting us to contact our respective units. That message flashed across the screen at approximately 10:30 PM Saudi time, which equated to nine hours ahead of our home base in the states, or ten hours ahead during daylight savings time.

Following the notification, we convened with our unit and were briefed that for the next 48 hours, we were to wear our uniforms exclusively, with no allowance for civilian attire.

On October 7th, the United States initiated airstrikes on Afghanistan. I found myself stationed at the weapon site, with missile launchers primed and prepared for action. Over the subsequent four days and nights, we witnessed a relentless barrage of airstrikes, with the onslaught persisting until October 20th.

As time progressed, numerous threats and accusations were exchanged, yet none materialized into significant actions until the conclusion of 2002. It was then that the United States commenced a buildup in Kuwait, laying the groundwork for potential conflict with Iraq should diplomatic demands remain unmet.

Despite our scheduled departure from Saudi Arabia in November 2001, the weight of the events we had experienced continued to linger heavily upon us all.

Second West Asia Tour

Iraq (2003)

Upon returning to our home station from Saudi Arabia and preceding the military buildup in Kuwait at the close of 2002, I underwent a permanent change of station move to Germany in July 2002. As the year drew to a close, tensions escalated in the conflict with Iraq. Units commenced deploying to Kuwait in anticipation of impending war.

Amidst the growing seriousness of the situation, my unit grappled with persistent rumors of potential deployment, yet no official orders were forthcoming. While units from both the states and abroad began receiving deployment directives into the new year of 2003, my own unit remained in a holding pattern, awaiting further instructions.

The long-awaited day finally arrived when we received our deployment orders. "Elements of the 6th Battalion 52nd Air Defense Artillery have received deployment orders," was the response from the Public Affairs Office when queried about our battalion's deployment to the war zone.

Initially, only one unit from my battalion was slated for deployment. However, a week later, the entire battalion was put on alert, signaling our imminent deployment. The official confirmation of our pending deployment came two days after my birthday, on February 22, 2003. Our call to action had finally arrived.

Within a week, the entire Battalion had loaded its equipment onto rails and ships in preparation for departure to Kuwait. Less than a month later, those of us scheduled for deployment were en route to the impending war in Iraq. While none of us desired to enter into conflict, we were all eager to fulfill our duty.

Approximately five hundred soldiers from Fifth Corps had received deployment orders to report to the Central Command area of responsibility as part of the buildup preceding the imminent war in Iraq. Soldiers were instructed to deploy between February and March of 2003. Fifth Corps forces were slated to offer unique capabilities to the Central Command Commander. However, the duration of the deployment had not been specified at the time the orders were issued.

On February 22, 2003, my unit received orders for a temporary change of station; our home base was in Germany. Those orders directed the movement of personnel and equipment from Germany to the US area of responsibility in Kuwait, where we were tasked with providing ongoing support to Operation Enduring Freedom and future operations.

The orders stipulated that our deployment would not exceed 365 days or extend beyond the completion of the mission, unless otherwise directed by the Commander. That provided a measure of relief, knowing that there was at least some semblance of an end date to our deployment. The decision was made for us to move as a cohesive unit, eliminating individual movements. That endeavor would be monumental, requiring meticulous preparation to relocate an entire unit, comprising personnel, gear, supplies, and equipment, to a distant location. Essentially, it meant transferring an entire base of operations to a new and remote destination. The logistical intricacies involved in such a massive undertaking were staggering.

Numerous members of the leadership were relatively inexperienced in that aspect of military operations and conflict. While some had prior experience in conflict zones, they hadn't necessarily been in key planning roles for such operations. Consequently, many of us had to lean heavily on assistance from various sources.

We extensively reached out to our contacts, conducted countless internal meetings to strategize the move, and meticulously mapped out a multitude of scenarios to guarantee the success of the operation. Our primary goal was to ensure that we had all the necessary resources and provisions at our new location.

As we readied ourselves for deployment, we also had to ensure that our home base would be adequately staffed with personnel to maintain communication with the deployed unit and to continue tracking and reporting on the status of individual soldiers. Not everyone from our unit would deploy to the war region. Our days grew increasingly busy and stretched as we made preparations for our imminent move.

Given the uncertainty surrounding the length of our deployment, we were authorized to store personal effects and arrange for the storage of our vehicles if necessary. As our departure date drew nearer, the reality of heading into combat became all too palpable. We were issued genuine combat gear, no longer training gear; that included actual biological gear, body armor, unused protective masks, unused Kevlar helmets, and live ammunition.

My birthday month in 2003 would certainly be etched in my memory, as it marked the time when I prepared to deploy into a combat zone. Such a significant moment was bound to leave a lasting impression.

Training prepared us for combat!

See book eight for the full Iraq story.

Third West Asia Tour

Kuwait (2007)

Returning to Kuwait years after my initial visit held a stark contrast. During my first trip, Kuwait served as a temporary rallying point where we awaited orders to move into Iraq, receiving equipment and gearing up for the mission ahead. However, in 2007, I found myself attached to a unit assigned to the region for an extended period. That time, it was not a brief layover; instead, I would spend months serving in the area, witnessing its dynamics and experiencing its challenges firsthand.

Almost four years to the day from my initial deployment to Kuwait, I found myself returning as part of an attachment crew with another unit. The decision came after extensive deliberation among leaders in the Brigade. The Battalion tasked with the mission lacked a sufficient number of certified tactical missile crews, prompting them to requisition crews from other units within the Brigade.

My crew was selected for deployment primarily due to our training, seniority, and availability at the time. After much speculation about whether or not we would be deployed, my assistant, our communication crew member, and I were ultimately placed on orders to support the mission in Kuwait. And so, back to the combat zone I would go.

On February 15, 2007, I received orders to deploy to the Central Command area of responsibility once again, that time to participate in Operation Enduring Freedom. Unlike my previous deployment, I was not involved in the unit's decision-making processes. Instead, my crew's role was solely focused on monitoring the skies for potential threats.

It's important to note that that deployment was not a permanent change of station; rather, it was temporary. We were attached to the deployed unit, but still permanently assigned to our home unit at Fort Bliss. The distinction underscored the temporary nature of our deployment and reinforced our ties to our home base.

The return trip to the combat zone brought about slightly improved living conditions, although far from luxurious. It wasn't a five-star or even a one-star situation, but I had no room to complain. It was a noticeable step up from the harsh conditions of living in the desert with no shelter except for our vehicles, as we experienced during our time in Iraq.

Tents with AC!

My own little living space!

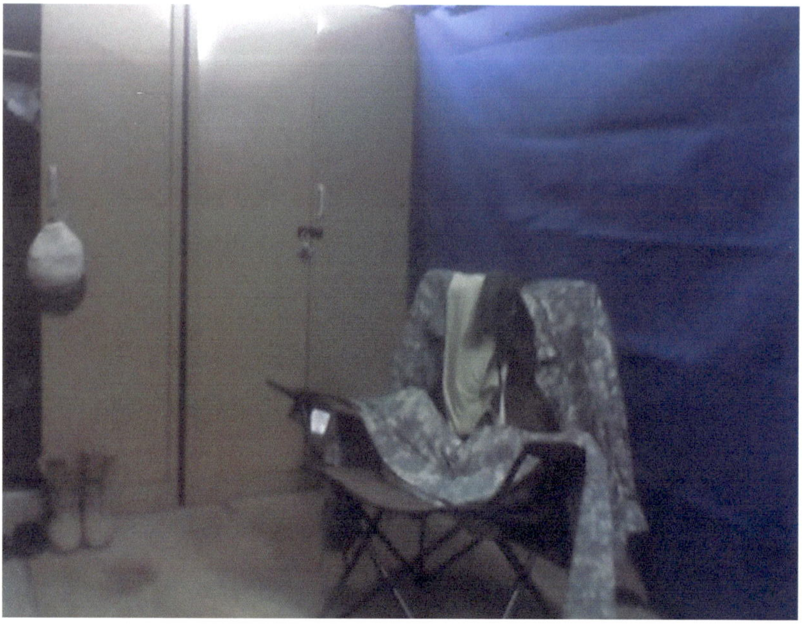

Experienced at making limited spacing work!

Later on, we upgraded from tent living to tin buildings, and it felt like a significant improvement. We couldn't help but liken ourselves to the Jeffersons, "movin' on up" to a better standard of living.

In Kuwait, we were introduced to amenities that were either absent or unfamiliar during our first trip and were certainly lacking during our time in Iraq. On-site, we had access to a barbecue area, an outdoor theater, and a basketball court, providing us with opportunities for relaxation and recreation amidst the demands of our deployment.

Additionally, there was a community center that served as a hub for various activities and gatherings. It housed a lounge area, a meeting room, and other amenities, providing a comfortable and communal space for relaxation and social interaction amidst the demands of our deployment.

In addition to the community center, there was a dining facility that offered the absolute best food, and a post exchange area that provided a range of amenities including a beauty salon, cleaners, and eateries such as Starbucks, Pizza Hut, Subway, KFC, and others. Kuwait the second time around proved to be a far cry from the first trip and a stark contrast to Iraq.

Furthermore, there were showers and laundry facilities available, which felt like a luxury compared to the primitive conditions we had experienced before. Gone were the days of showering in a wooden box or doing laundry in a plastic bag; the newfound amenities truly spoiled us.

Once settled into our living quarters, my assistant, communication technician, and I would convene with the leadership to receive a briefing on the unit's mission and to understand our specific duties.

The unit's primary air defense responsibilities centered around defending against missile threats targeting US interests in Kuwait. My crew, assigned to the wee morning shift, had the crucial task of monitoring the skies for potential threats and being ready to engage if necessary.

Remaining attached to the unit for several months, we continued our early morning shift diligently. However, my return to Kuwait marked a significant milestone due to my extensive tactical experience. I was deployed to serve as a tactical director in support of ongoing combat operations in the region.

My duties encompassed coordinating the activities of all task force elements, which involved correlating tracks observed on the monitor, resolving identity conflicts, establishing engagement priorities for firing units, and monitoring missile engagements. That responsibility demanded long hours and unwavering vigilance. The stress of the role began to take its toll on myself and my crew as we navigated through the challenges of combat operations.

Multitasking, the stress was real!

After months of performing our duties diligently, the news of our unit's request for our return to Fort Bliss brought a sense of relief. Our unit was slated to rotate to Korea, and they required all their crews. While it was probable that we would be performing the same job in Korea, we didn't anticipate the same tense pressure that we experienced in Kuwait.

My crew and I wrapped up our tour of duty in Kuwait, receiving thanks for our support to the unit. We packed our belongings and headed back stateside to prepare for yet another deployment.

Made in the USA
Las Vegas, NV
07 November 2026

57669923R00026